ZEGE Inc

Electro Magnetic Induction

Zero Emission Global Energy LLC

ELIMINATING FUELS IN EVERYTHING EVERY WHERE

VIDEO & PDF doc. documentation is ICON loaded on your ZEGE Inc. Tablet

ZEGE-SEA-FUELLESS

Announces it is buying 100 MITSUBISHI International Corporation cargo and container ships. They will be retrofitted with all electric self sustaining energy systems from EMI-ZEGE using Phoenix Energy of Nevada's EMI technology, a zero-emission, **zero-fuel** and zero emission, unlimited mileage technology for $200 billion.

Mitsubishi Heavy Industries will soon be ZEGE-PENV-EMI tech Manufacturing, Licensed & Certified

ZEGE-AIR-FUELESS

ZEGE-AIR to order $2Trillion, 10,000 MRJ MITSUBISHI turboprop, turbofan planes and jets 20 million seat/flts capacity a day. City to City to Suburb MRJ Commuter **short-mid-range to unlimited distances. Short, mid-range to long range flights will use Phoenix Energy 100Mwe electric steam turbine, self sustaining generators, manufacturing on board, as needed, hydrogen gas fuel from H2O Electrolysis using water in the fuel tanks instead of jet fuel.**

Global long distance routes can be scheduled or chartered VIP Groups all using EMI PENV ZEGE hydrogen fueled retrofitted planes with only minor changes to the Jet and turbines.

ZEGE-Airlines shall use mainly Mitsubishi planes and Phoenix emi energy tech and will be the largest airline in the world.

Z EGE will fly the very first 3 purchased MRJ prop planes in 10 nonstop **fuelless** circumnavigations of the Globe in 780 hours to demonstrate just what Phoenix Energy of Nevada's EMI technology can achieved in almost any large moving transportation vehicle used on land, air or sea.

The first stage of our fuelless, unlimited distance flight goal, is to use the Mitsubishi MRJ counter rotating, or multi blade propeller planes converted to electric motor driven propulsion instead of diesel generators. The 100MWe EMI Phoenix Energy electricity systems will give us enough power to more than equal current rotational energy in the propeller system.

The Prop plane retro fit we will go direct drive with large proven Mitsubishi electric elevator high torque motors to drive the multi-blade propellers. Little changes for a no fuel and longer range results.

Z EGE-Mitsubishi-Phoenix, goal is to retrofit many types jets to NO FUEL and UNLIMITED MILEAGE takes on this bold new horizon with enthusiasm. We trust in the simple facts of the beauty of heating high strength and high temperature metals with Electromagnetic Induction which has been used in the US Steel industry for over 50 years. Heating high temperature metals inside the turbine combustion chamber and beyond is child's play for ZEGE Inc's Patent Pending EMI retrofitted turbine technology.

The turbine combustion chamber may have to be elongated or/and widened to accommodate the new induction heated 3,500 degree metals surrounded by ceramics for high heat protection in the engine turbine shroud, that mimic and replace the jet fuel combustion, but I doubt that. I am confident our changes will be minor to the already certified engines. Performance tests shall prove what differences should be made.

Engineering physics will prove with testing just how hot and what minor metal placements to increase the superheating inside the fuel chamber of the turbine from 3,500 degrees fahrenheit usual temperatures or even higher if needed, by a proprietary Mitsubishi or the GE ceramic housing protections embedded with the copper magnetic coils. We can continue to superheat the compressed air using more ceramic protection for the EMI coils embedded further down the turbine finned chamber inside the shroud heating h super high temperature metals configured to heat the turbine air continually down the length of the existing thrust exhaust distributor as it travels through the turbo fins to acceptable temperatures exactly as liquid diesel or hydrogen gas does.

Exactly is the operative word here. If we do that we are golden. NO FUEL, for unlimited distances flights will have been achieved using all electricity ZEGE Inc. patented EMI turbine internal retrofits further powered by the Phoenix Energy of Nevada EMI 100MWe on board power plants using electricity from an Electro Magnetic Induction, high pressure, closed loop, steam turbine systems in the lower deck of the aircraft. These new revolutionary electric power plants are work horses that use no fuels themselves. Phoenix Energy of Nevada, LLC of Carson City, Nevada's new technology is the only reason any of this is even possible to achieve. This will be a major wonder and aviation game changer cutting fuel cost to zero and allowing unlimited mileage flight.

Changing from jet fuel to hydrogen is a great thing but changing fuels to NO FUELS is aviation INSANITY, but that is what we are able to do with internal changes to the turbine or jet engine fuel burning chamber where air has been pre compressed then super heated with jet fuel or in the hydrogen verion to 3,500 degrees fahrenheit. With electromagnetic Induction we can super heat high temperature metals to 7,000 degrees if need be, to mimic the thrust parameters jet fuel currently give us using Phoenix Energy tech of EMI instead.

Changing out the turbo and jet engines to accept, on board manufactured, as needed hydrogen gas to fire the turbines, allows us to go zero emission hydrogen manufactured on board, and Long Range, BUT NOT, unlimited distances. Unlimited mileage takes another bold ZEGE EMI Patent Pending with PENV tech retrofit. We cannot load enough water to manufacture enough hydrogen fuel on this trip, so we will lean on the onboard 100MWE Phoenix Energy of Nevada's EMI tech steam turbine to give us electricity output right where the diesel of Hydrogen fuels heat the turbine air to create the very same heat energy transfer to the turbine or jet systems, to come to our rescue for this ZEGE-EMI undertaking.

Annual Jet, Prop or TurboProp or Turbofan Revenue $10-30 trillion.

ZEGE-EMI Patent Pending aircraft turbine changes, multiple manufacturing global licensing agreements to achieve no fuel, unlimited mileage $5+ trillion.

ZEGE-AIR-TOURS TOURS TOURS

24 hour -30 day global tours

If you have to ask the price you can't afford one.

VIP Services include: overnight luxury hotel, meals, and ground transportation.

We offer group and First Class or Coach Accommodations pricing.

24 hour anywhere in the world, 1, 2 week and 30 days tours

ZEGE-Mitsubishi –Phoenix -MRJ Prop and Turbo Prop , Turbo Fan and Jets

Annual revenue serving 50 million/ 100,000/day seats MRJ $1 trillion.

ZEGE-AIR Propulsion

Zero Emission-Fuelless

- **ZEGE-AIR Propulsion will retrofit the** Turboprop engines to be hydrogen gas powered
- ZEGE-EMI PENV all electric
- Top secret patented glowing red titanium 5,000 degree jet turbine shrouds and turbine fans and turbine compressive blades heated super hot by electro magnetic induction currents totally eliminating diesel with or without hydrogen gas fed.

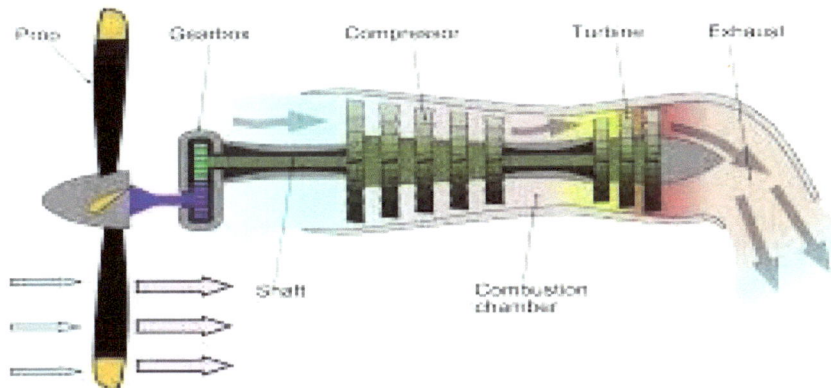

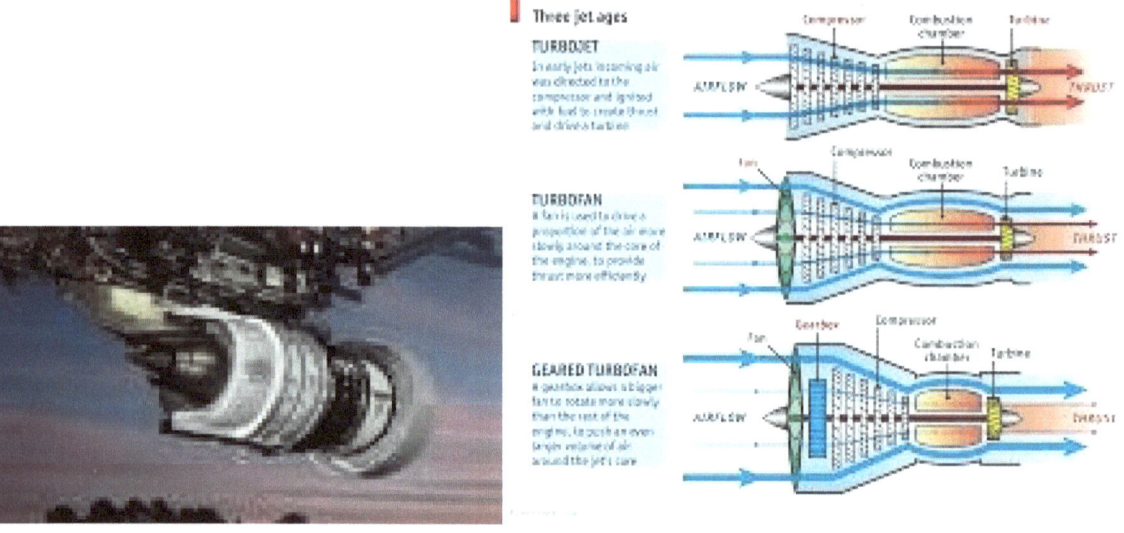

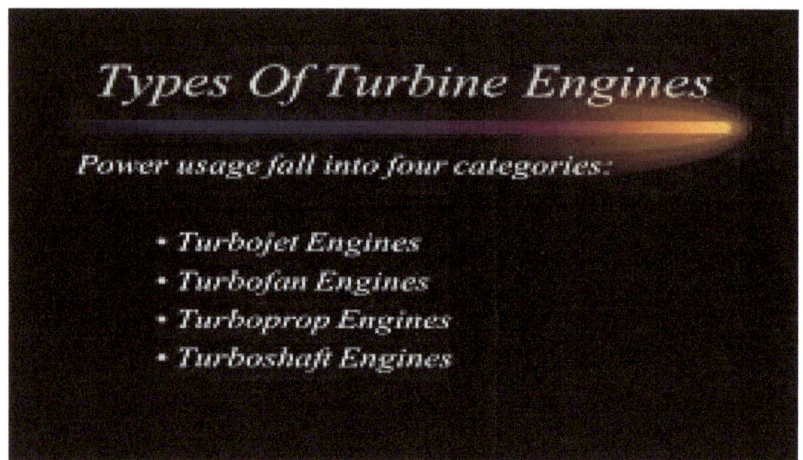

Types Of Turbine Engines

Power usage fall into four categories:

- Turbojet Engines
- Turbofan Engines
- Turboprop Engines
- Turboshaft Engines

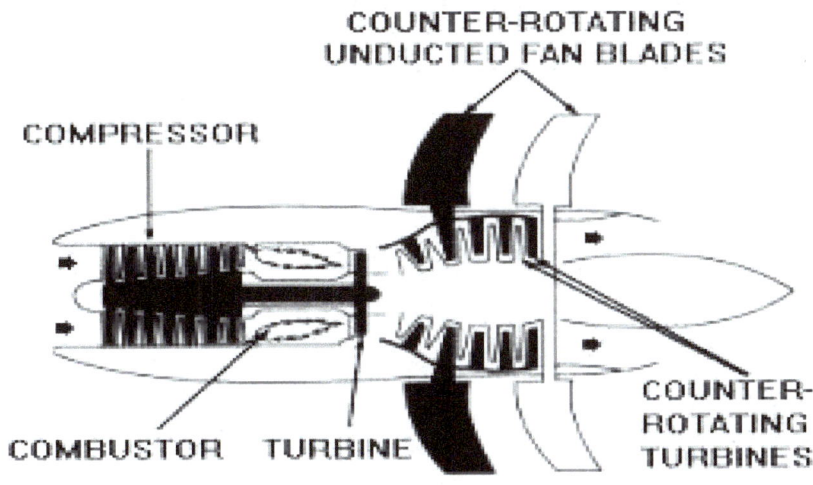

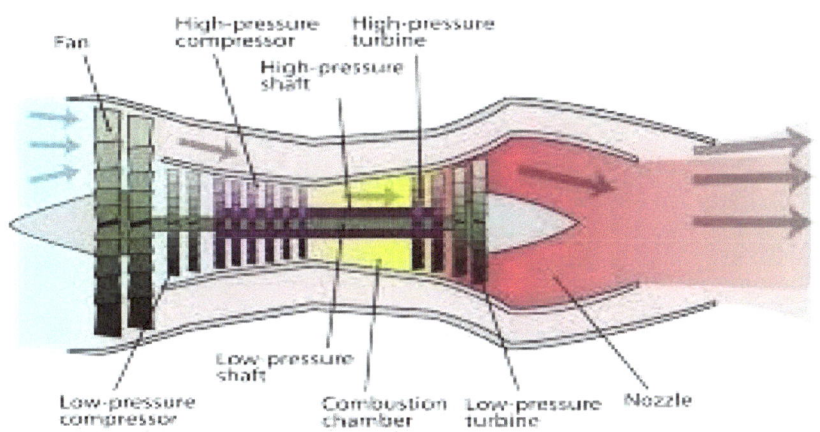

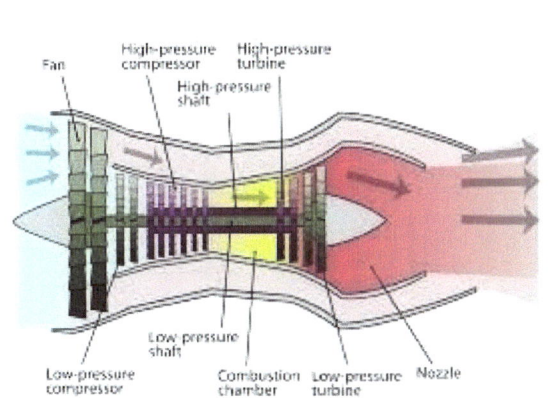

In the ZEGE EMI TOP

SECRET engines all the blades and the shroud will be super temperature materials including the shroud inner lining to be electromagnetic induction heat transfer structures super glowing red hot to 3,500-7,000 degrees. Hydrogen gas will be added instead of jet fuel and the fuel tanks will be filled with water for electrolysis hydrogen gas manufactured by the on board ZEGE-PENV- EMI electrical high pressure steam turbine.

In ZEGE- emi retro fitted prop and turboprop configurations direct drive electric motors maybe replacing the diesel fuel for propulsion. IN our other retrofits in prop, turboprops, turbofans and jet engines hydrogen replaces diesel.

WE are all about eliminating all the fuel, everywhere here at ZEGE-emi.

ZEGE-emi

TRAIN-LOCO _{ZEGE}

EMI, MODIFYING TRAINS by eliminating the fuels.

That's it?

Yes!   Where not LOCO!

Fast!   Simple!   Easy!

We're just Brilliant!

DIESEL?

DIESEL?

DIESEL?

Stinking Diesel?

Say, you are not still using diesel, in your trains are you?

You don't need no stinking diesel.

Go emi in your Locomotives and stop using fuels.

Call 808-397-0344: Jeffrey

916-517-7007: Chad

ZEGE-FUELLESS-TRAIN

We retro fit your diesel locomotive by Embedding our 100MWe ZEGE-EMI-PENV electric steam turbine system making your Engine an all electric propulsion drive locomotive. All Electric Locomotives will be a fast, clean retrofit, creating a huge engine to drive revenues, by **eliminating fuels.**

Go Coast to Coast with us as we change the way you think and fuel trains.

No More Diesel.

More Train deals needing a ZEGE-PENV-EMI no fuel fix.

High SPEED China

ZEGE-AIR TOURS

Alaska and Mt. Ararat

Hanoi

Serengeti Heard Migration Season

Victoria Falls, Afrika

Japan

Sand Castles — Malta

Ireland

Come tour the world up close like you could never do from 50,000 ft high jumbo in a Mitsubishi- MRJ TurboProp long distance fuelless ZEGE-Phoenix-EMI powered plane.

ZEGE-FUELLES-SEA

An ocean going cruise liner division has been added to Zege-Sea using the new Phoenix Energy's zero emission technology for unlimited no fuel mileage. A 25 cruise liner ship, $75 billion order with MITSUBISHI Int. Corp. These ships will be fitted with the revolutionary technology of Phoenix Energy of Nevada, LLC, allowing totally zero pollution EMI systems which treats sewage, desalinates sea water, heats the cabins and hot water, as well, fires up the kitchen stoves while lighting the ship and powering the controls and propulsion system all while using no fuel. Passenger pricing will reflect the mid-lower end ranges (bargain basement) below Jet travel. The low priced tickets are mainly due to the revolutionary Phoenix EMI technology that has eliminated the cost of fuel allowing unlimited mileage.

Other emi-ZEGE, LLC divisions and orders placed with MITSUBISHI of Japan financed by the major Japanese banks as well as Saudi Arabia private venture funds, and German banking, included is French PNB Paribas and Chinese world's largest bank CICB are: (all equipment is retrofitted with PENV EMI no fuel zero emission technology)

ZEGE-emi Sales

Contract Sales Projects, Manufacturing Certification and Licensing, PPA, Lease, Design, Build, Own & Manage.

ZEGE Contracts, Manufacturing Contracts, sub parts manufacturing , Licensing etc Phoenix Energy per EMI unit fees, Oversight and special usage engineering and tech oversight. $1-$10 trillion

ZEGE emi- in house Global Sales/ Contract/Manufacturing/$$$Comissions/ Operations/ budget… Private

ZEGE-emi HOME Depot 30Mwe ON SITE units 1500 stores electric and heating $75 billion

ZEGE-emi Lowes 1500 Stores 30Mwe units $75 billion

ZEGE-emi Walmart-Sams Club 1500 Stores ON-SITE 100Mwe power electric, heat & hot water $120 million each…total $75 billion

ZEGE Walgreen-Rite Aid 7,000 Stores 30Mwe On-SITE $50m ea. $180 billion

ZEGE-emi CVS-Longs etc. 4,000 Stores 30 Mwe On SITE $50m ea. $90 billion

ZEGE-emi Groceries $120 million each 100Mwe

ZEGE-inc Phoenix Energy 600 CLOSED COAL Purchase & PENV EMI Retrofit Project $100 billion paid for by ZEGE inc from the Mitsubishi Licensing Agreement.

ZEGE-inc Bank Of Hawaii High Rise one 30Mwe unit $40 million. These zege-penv-emi units generate zero energy cost with zero emission, are 10 times too large so we are configuring the PPA lease agreements to sell all excess continual electricity back to the HECO Grid for 10-12 cents/Kwe cash profits of over $25 million yearly to the bank minus ZEGE COST PLUS fees.

ZEGE-inc First Hawaiian Bank High Rise one 30Mwe unit $40 million PPA Lease agreement bank profits with no electricity costs, generating bank HECO GRID paid profits of $25 million minus ZEGE COST PLUS fees.

ZEGE-inc plans are to embed Phoenix Energy of Nevada's Electromagnetic Induction technology into 1-300 00 ON-SITE business high electric use businesses and Office, Apartment and Condo High Rise projects included are Malls as well as Sand Island Sewage treatment and Hawaii Water treatment saving these two entities and Hawaii residents $tens of $billions in future FED mandated costs across Hawaii.

Most opportunities for ZEGE financed projects will be in Honolulu and Waikiki creating a 100% clean HECO Grid solution. This will provide ZEGE PENV EMI and HECO or and Hawaiian Gas to start to bring clean renewable fuels to the transportation needs of Hawaii which is 2 times the energy demand as the HECO electric grid can provide until now with these ON-SITE installations.

ZEGE_PENV-EMI tech Hydrogen conversion from the excess electricity can be manufactured and stored for future grid or transportation energy needs. A major hydrogen North South trunk line funded and managed by Hawaiian Gas Company will allow many clean energy hydrogen utilization projects eliminating current expensive electric pricing for those who switch from grid electricity ON-SITE independent energy hydrogen consumed new tech equipment. See our Hawaii Energy project full disclosure report below in full documentation and detail.

Hawaii DEREGULATION

Of electricity production at Power Plants from Grid distribution under Federal Energy Laws and PURPA Regulations.

ZEGE is asking Hawaii PUC and Legislature to deregulate the HECO GRID from HECO Power Plants and sell off one or the other to create new clean renewable Fast Track implementation with local BOH and FHB financing $10 billion for the current 2500 Mwe total Hawaiian electrical consumption HECO operations.

ZEGE/PENV. HECO runs Grid…operates and updates plants with retrofits EMI PENV 30 units 100Mwe $120 ml each. 2 new power plants $120ml. ea, replace turbines $300ml total COST PLUS Cost: $1.2 bl Plus 8cents Kwe 1= $1.6 billion Kwe revenue first year to pay off $1.2 bl BOH CEO Peter Ho/FHB CEO Robert Harrison loan. year 2 & 3 drop Kwe rate to 1 cent/Kwe= $200 million a year paid to ZEGE PPA. HECO HECO customer savings/yr. $300Million-$1 billion. 30 year savings $10-$30 billion. Year 4 HECO plants handed over with PUC authority to **Hawaiian Renewable Energy Alliance** who will take over operations with their own management for long term local operations with low costs.
ZEGE-emi Air Freight Prop Planes unlimited miles $220 million ea.

ZEGE-emi Trucking…10,000 Mitsubishi TUSO long haul trucks no fuel…wholly owned and self funded. $5 trillion.

ZEGE-emi Bus lines…10,000 Mitsubishi TUSO executive commuter buses no fuel wholly owned and self funded $5 trillion

ZEGE-emi 10 million auto hydrogen fueled: $500 billion. Japan Hydrogen VISION

ZEGE-emi 10 million auto electric powered: $400 billion   Japan VISION 2020

ZEGE-emi   5,000 Global interstate truck **fueling stations**: diesel/hydrogen/electric /restaurants/inns/convenience stores. Build-own-operate.$500 billion self funded.

ZEGE-emi LOCAL Kwe POWER $3 trillion new and retrofitted electric plants fully owned or PPA agreements self funded.

ZEGE-emi-ON SITE Kwe Power 30,000… 100Mwe units $100-$500million each units per project build to own/ operate or PPA agreements or sell to customers. $3 trillion

ZEGE-emi Syntech Diesel from emi derived hydrogen H2CO2Fe

ZEGE-emi Syntech JET Fuel from emi derived hydrogen

ZEGE-emi HYDROGEN Transportation. Heating, Hot water, Syntech diesel and jet fuels. EMI Hydrogen GRID Energy Storage Systems.

ZEGE-emi Snow Melt Removal High Pressure Steam Trucks driving at 40-60mph melt snow with on board PENV EMI systems to clear snow from roads and airports using 30-50MWE ZEGE-PNEV-EMI 1,000-10,000 units Globally $40 million each unit needs no fuel for operation or driving all electric self sustaining manufactured IN-SITE from Phoenix Energy electromagnetic Induction tech……total $40-$400 billion

ZEGE-emi-RETROFITTING POWER PLANTS 20,000- 50,000 units Global sales $15million/100MWe each unit $300-$750 billion

ZEGE-emi Groceries, fruit and nut tree farms grains and vegetables, farms, herds, chickens, eggs, milking cows, cattle herds, fish ponds, all locally sourced and produced. Self-sustaining projects will be developed with emi technology. Once they are successfully embedded with emi clean renewable tech and robustly profitable we may sell them to like minded business companies. This will allow further projects to be self sustaining as well. We need what athletes call cascading events, building several and selling and building again and again all around the world is just that a cascading economic event that is self sustaining and duplicating. Each community will be funded to provide for area needs and manufacture disaster provisions for millions for a 10 years disaster relief event food and seed storage shelters.

ZEGE-emi FARMING Saudi Arabia...Greening the Desert...Creating ten shipping canals leading to a huge manmade Destination sea with super cooles fresh water desalinating fresh water towers. This project will sustain many new decentralized suburbs all zero fuel and zero emission self sustaining Nut and Fruit, and grain farming with robust employment from large food and tree seed growing and drying operations able to feed billions from storage systems against global catastrophes lasting 10-20 years. $5 trillion annual budget.

ZEGE-DIVISIONS

ZEGE-emi Phoenix Energy of Nevada
ZEGE FUELLESS Research
ZEGE-emi MITSUBISHI Int Corp.
ZEGE-emi Mitsubishi Power
ZEGE-emi Mitsubishi Shipyard
ZEGE-emi GE
ZEGE-emi US
ZEGE-emi Japan
ZEGE-emi India
ZEGE-emi China
ZEGE-emi Siemens
ZEGE-emi Inductotherm Corp.
ZEGE-emi FLOUR
ZEGE-emi Wilcock & Babcock
ZEGE-emi Rolls Royce
ZEGE-emi Boeing
ZEGE-emi BROWN UPS
ZEGE-emi FEDEX AIR GLOBAL Integration Systems
ZEGE-emi Nissan
ZEGE-emi Toyota
ZEGE-emi Honda
ZEGE-emi USA
ZEGE-emi Japan

ZEGE-emi Saudi Arabia ten seawater shipping Canals far into the desert leading to a huge new DEEP SEA with low energy consuming Desalination Condensing tower systems to sustain new desert suburban communities. Thousands of 100MWE ZEGE PENV EMI decentralized ON-SITE interconnected independent power plants.

ZEGE-emi Germany The PENV EMI 40 cent/KWE FIX $3trillion
ZEGE-emi Britain $1 trillion
ZEGE-emi France $1 trillion
ZEGE-emi China $10 trillion
ZEGE-emi India $10 trillion
ZEGE emi Canada $500 billion
ZEGE –emi Bombardier
ZEGE-emi Hawaii
ZEGE-emi Pacific Rim
ZEGE-emi Middle East
ZEGE-emi S. Korea
ZEGE-emi South America

EMI Zero Emission Global Energy, LLC (ZEGE) is inviting several respected banks to be on its financial team as our paid corporate advisors and investment bankers to handle our Sovereign Fund and commercial loans, lines of credit, financing instruments, purchases/sales dedicated contracts, licensing agreements and various dedicated escrow accounts with our multiple revenue streams of $10-$20 trillion dollars

- Bank of Tokyo –Mitsubishi UFJ
- Sumitomo Mitsui Bank Corp.
- Mizuho Bank Ltd Japan
- Japan POST Bank Co. Ltd
- Japan Mishkin Banks
- Saudi Arabia Vision 2030
- German
- French
- British
- BOH Bank of Hawaii
- FHB First Hawaiian Bank
- PNB Paribas
- CICB
- AIIB

(ElectroMagnetic Induction)   EMI Zero Emission Global Energy LLC

EMI-ZEGE LLC

ZEGE LLC

ZEGE

EMI ZEGE LLC BANKING ALLIANCE

Shall include the previously mentioned banks and these many other Japanese banks have been invited to participate in funding the Mitsubishi ZEGE PENV EMI Manufacturng Licensing Agreement with EMI ZEGE LLC to embed Phoenix Energy of Nevada, LLC EMI technology in products that are mainly funded through financing, leasing, loans and lines of credit with many needed escrow accounts by Japan to foster great growth and general employment and wage increases throughout Japan in a swift robust planned structured way driven by their financing exact schedules of contract manufacturing orders with Mitsubishi International Corporation's many Groups and Her selected sub contracted manufacturers.

Japan Bank of International Cooperation

Aichi Bank

Bank of Japan

Chiba Bank

77 Bank

Bank of Fukuoka

Shizuoka Bank

Yamaguchi Bank

BOY Bank of Yokohama

Bank of Nagoya

Bank of SAGA

Nanto Bank

Nomura Holdings

Norinchukin Bank COOP

Oita Bank

RESONA Bank

RESONA Group

The EMI ZEGE LLC Banking ALLIANCE shall invite each party to invest in many parts of the Financing Instruments and in so doing an initial investment deposit to cover the Manufacturing Licensing Agreements between ZEGE LLC-PENV and Mitsubishi of $260 billion and the Opening orders to manufacture of $2 trillion.

Each investing bank is asked to commit to 6 times 10% of their assets. In return we shall return many revenue streams based on percentages of the amounts each bank agrees to invest in the EMI ZEGE, LLC Banking ALLIANCE.

The more that any bank invests, and or rolls over to re-invest as we proceed shall determine what portion of the many revenue streams that bank and any and all other participating investors or investing banks shall receive. Documents spelling out what minimum investments are expected to be invested by each party and how all the needed instruments shall be agreed upon through legal contracts. The Investment Operating sum shall be the basis of EMI-ZEGE, LLC Banking ALLIANCE Line of Credits, Loan, Escrow Accounts, Leasing Financing, Research & Development, Projects Funding, ON-SITE Project Funding, Marketing, Sales, Manufacturing Costs, Power plant project retrofit and new plant funding, Hydrogen Projects manufacture of Energy funding, EMI-ZEGE PENV-EMI Licensing Japan Mitsubishi covering its subcontratos in Japan, predominantly and of manufacturing in US and Europe as determined by ZEGE-PENV and MItsubishi to favor Mitsubishi and Japan.

China Manufacturing License Agreements as well, Saudi Arabia/Middle East, and India Manufacturing Licenses shall be at the discretion of EMI-ZEGE and Phoenix Energy of Nevada or PENV solely. And will not change unless EMI-ZEGE, LLC and Phoenix Energy of Nevada. LLC stipulates in an addendum to the Banking Alliance to agree to do otherwise by folding in those Licensing areas as financing partners to the initial EMI-ZEGE, LLC Banking Alliance. Let all parties know that EMI-ZEGE, LLC and not PENV is institututing and making all contractual agreements with Phoenix Energy of Nevada, LLC and MItsubishi to manufacture and make orders and payments for all products and projects under it's own agreements with Phoenix Energy when embedding PENV-EMI technology into any system or as stand alone in power plants or retrofitted power plants or any other transportation vehicle being of Land, Air or SAE, Trains, Planes and Ships, etc.

The EMI-ZEGE, LLC Banking ALLIANCE shall also engage in Social-Crowdfunding for levels of investing in OUR ON-SITE and Independent Power plant and hydrogen energy manufacturing projects, lowering anyone's individual electricity bills globally with 1-2 and 3 year annual proven electricity bill total costs that they will invest in dedicated power plants that we will install around the world to generate incomes from Power Companies in PPA excess electricity production by us to not only benefit the ZEGE Banking ALLIANCE but these individual investors as well which lowers their electric bill depending on their investment amounts and gives the larger investors a revenue stream from investing in EMI technology. If you see Global advertising or tweets about ZEGE paying customers electricity bills you know it is our program.

ZEGE-MITSUBISHI-PENV

Partnership Banking Strategies

We need the few large banks to keep us fluid with cash at the beginning with a large line of credit ongoing. Most projects will pay for themselves in less than one year. I will pay off all loans with profits to the banks. I do not want the banks to hesitate in funding all our requests for cash, credit and financial advising.

ZEGE-emi will bank with our funding partners, we could after a while become a self funding enterprise and create our own banking entity. But my goal is to have many financial instruments for our customers and divisions to utilize. I have followed GE Financial and their biggest mistake was that they never wanted to fund larger than $100,000 equipment projects. Now for us that would be a death knoll.

Every project we are undertaking is $millions and $billions sometimes $trillions. Who in their right mind wouldn't know they needed to help their lenders grow super large enough to roll with the need.

Most equipment sales will be prepaid prior to manufacture. We will ask our banking partners to finance our customers purchases with financing loans. This way of dealing with huge monetary demands on ZEGE GLOBAL FINANCE is designed to keep money flows and profits coming to us with quick turnovers, and is to create as many multiple revenue streams the banks. Any enterprise to be successful must be able to have an exponential growth possibility if it does not grow faster and faster it cannot be sustained and it will quickly fail. I hope the banks ride this to great wealth creation as we infect the rest of the world with EMI technology too.

All the big rich men of wealth have divested of oil and gas stock banking on new clean technology. Just recently, Rockefeller quit ExxonMobil, completely divesting. Google too. Warren Buffett investing in wind for a PA of only 3 cents/KWe soon after, Microsoft, Oracle and Apple have all followed suit. Coal companies are all bankrupting to natural gas's low prices. The US renewable investment industry is now $500 billion a year.

Saudi Arabia Deputy Crown Prince Mohammed bin Salman now Finance Minister said oil is dead to him. He is tasked to fulfill the VISION 2030 plan to curtail the kingdom's addiction to oil. Deputy Crown Prince Mohammed Salma is selling part of ARAMCO 5% for $3trillion to invest in new clean renewable. Mr. Salman likes the Phoenix Energy technology emi ZEGE is selling at COST Plus 2 cent/Kwe electricity. Salman has a contract for wind at 6 cents/Kwe. I have invited him to join with the United Arab Emirate whom are huge renewable funders in the Middle East and into Egypt and Turkey, to join the large Japanese Banks with French PNB Paribas, China CICB the AIIB and German banks in funding ZEGE.

Even the smaller local Japanese Mishkin banks are flush with $trillions in cash, but have no manufacturing or large worker employment increasing manufacturing projects to stimulate the economy. Japan's MISHKIN Banks, by law can fund the manufacturing growth projects in Japan, just not overseas. But they can fund the ZEGE Mitsubishi manufacturing in Japan and its ZEGE orders subcontractors inside the country.

Let me introduce myself, I am Jeffrey Barrett, President of ZEGE, have my past financial projects to prove my readiness for this great venture. They are well known internationally.

- For instance, in the creation of the protocols for the AIIB.
- And the Yuan/Yen monetary Conversion Agreement.
- The success of the **FAST TRACK Economics** plans for both China and India will establish **ZEGE's Global Economic Qualifications** to attract great wealth to the project. Once we are established our new businesses will generate huge profits for all our banking partners and Manufacturers. Phoenix Energy EMI tech will allow ZEGE to dwarf the success of the very fast and long lasting growth and wealth creation of that great **English East India Company.**

Our shipping and airline divisions with trucking, Plant retrofits and new power plant build/own/operate…Interstate Fueling stations with all our products ZEGE manufactured, etc. etc. will be developed to be self funding with multiple incomes streams. These operational divisions are available to be sold off to duplicate them many times over in other countries etc.

ZEGE IS BUYING and EMBEDDING PENV EMI tech into equipment or to build ON-SITE energy systems creating cheap clean energy, then selling it to customers using fossil fuels to cut costs for them and eliminate their fossils fuel expenses. We are not just selling renewable energy equipment, but the importance of the economics of Cascading Growing long term wealth creation for us is this: we will not ask you to just listen and buy into our products.

Your companies have already bought and operate energy systems or are provided energy products by ongoing long term companies with true & tried old school technology. Why bother changing really why should ZEGE bother to wait for sales. We are in the ownership of high energy use products and systems. We will buy your Power Company, Office Tower, Grocery chain imbed Phoenix EMI tech into every energy use with the cost savings of needing no fuel and in transportation vehicles; Planes, Trains, Automobiles Cargo Ships and Cruise Liners we will give them all unlimited mileage capabilities.

ZEGE-emi   CREATING and BUILDING WHOLE NEW DECENTRALISED SUBURBAN COMMUNITIES

When we build whole suburbs we shall build in the ZEGE manufactured cheap energy products for heating, AC, water, water treatment, sewerage treatment, desalination, hydrogen fuels, cheap electricity, jet fuels and diesel all zero emission manufactured energy products communities need every day at work around town and at schools and workplaces in the suburbs. We do not import oil or natural gas we eliminate the need for it in every current use of this dirty expensive fossil fuel products that sap your paycheck. We are "DIRT cheaper and ZERO emissions cleaner.

Zero Emission Global Energy will own and operate the many companies you rely on now that cannot afford or want to bother to step up and make the necessary purchases and pay for them their selves. It is expensive to do. We will make products with the new technology already inside. We will them own operate and if one wants sell that enterprise which will compete with far lower costs and far cleaner emission operations. We are in the business to pay for the manufacture of the many energy intensive large infrastructure systems that everyone needs and uses, made better, faster, cleaner and cheaper.

Planes and ships with EMI will go ten times farther on less fuel. Let me restate that farther on NO FUEL. Now that is how we plan to drive growth through the roof year after year globally. We can then sell off these new expensive systems for prices you could never afford if you were to yourself have them manufactured. We will practically hand over our very profitable company divisions for enterprising business men to own and run so we can get on with bring clean renewable using ZEGE funded PENV Electromagnetic Induction technology patented by Phoenix Energy of Nevada whom we are looking forward to a long and profitable friendly partnership in everything we imbed their technology in.

We have one of the best Executive Salesmen running our EMI tech Phoenix Energy systems as do they and many other vendors they must and will select to represent them as independent businesses. We will sell 10's of millions of ZEGE PENV EMI systems to businesses who want to buy these new cheap, clean, energy intensive system to cut costs and eliminate pollution in their operations or to extend their equipments abilities beyond the manufacturer's designed operation. If you run a University and have huge energy cost and want to eliminate those costs call me, Jeffrey Barrett, ZEGE President to have an Executive Salesman or myself write up your order or answer your questions. We will gladly sell you what you need to get this done for you. What we will not do is cold, call you or knock on your door.

It is important to, prove the worthiness of the technology, buy putting it out in public use and in letting the media report on what astounding changes are available to anyone needing energy cost and pollution relief. We will bid on any and all energy projects around the world. We will manufacture millions of products before you even know you may want to experience the cost savings of our current old products or energy systems. When you come back from a fuel less Plane trip that went around the world without stopping for fuel you will be our best salesman spokesperson and we thank you in advance for that.

We will not be silenced. We will be loud and ever present in your everyday life as you get on a ZEGE/Phoenix tech powered bus, train, plane or cruise liner. When your water and sewer bill electricity bill drops through the roof, you just might be told EMI ZEGE PENV and Google these things to find out what we are doing. We plan to be very busy.

Please help us by asking your business or Power Company to install our technology. Ask your favorite Box store manager, hospital or University if they are totally EMI ZEGE PENV energy clean yet and if not why not? Ask your banker when you buy or build an office tower if it is EMI ZEGE PENV and if not make it so. Tell your government officials and PUC Chairman to let us come and PPA their Electric Grid power plants, water and sewer systems with EMI ZEGE PENV.

When you fly ask if the airliner is going to fly using clean EMI ZEGE PENV syntech manufactured locally jet fuel? Or buy short hop all electric prop planes with unlimited mileage. Hawaii is spending tens of billions to treat sewerage, They can buy and use EMI ZEGE PENV to clean it to drinking water levels for 1/100th the cost and make cheap clean electricity in the process.

If you work for a business or Corporation tell them they can build ON SITE EMI ZEGE PENV financed energy plants at 2 cents/ Kwe. Hey Power Companies watch out we are coming to take businesses away from you. We at ZEGE offer to buy your electric Power Plants and install MI ZEGE PENV technology systems and sell electricity back to you or your GRID Operators at 2cent/Kwe long term contract agreements or in a PPA with plant owners for 8 cents/Kwe or less for only one to two years to pay back our manufacturing/ installation/ operations/ financing and equipment costs.

To the cold winter snow shoveling Nor'easter type states and countries who heat with oil or gas think about asking your Electric Power Companies and Electric Grid operators to install our EMI ZEGE PENV technology so you can buy an electric car with home and work charging systems at low electricity prices> Then tell them you want to end your heating oil provider contract by heating your home cheaply and/or fuel your car, with low cost, high energy H2 hydrogen fuel converted by an in home electrolysis system.

Ask us about our **COST PLUS-Financing** products.

We are not just selling and manufacturing thru Japan Mitsubishi etc using Phoenix clean technology and Japanese etc financing. We are the customer making the orders for ourselves. We will be ensuring this technology is implemented globally as far and fast as possible. The Big Oil and Gas International Conglomerates with the old school Wall Street and Too big to fail Megabanks will stymie us if we wait for them to come to emi-ZEGE or Phoenix Energy to buy this technology.

That is a dead issue of their mindset to control the world's consumables and financing. We are doing an end run around them. Compared to designing the **FAST TRACK ECONOMICS for China and India's economic systems** and social and business agreements taught by Cyrus Feinstein in his MEGA tone **A Trillion Dollar Enterprise** and Fannie Mae concentration of wealth to reinvest and grow strategies with business court protocols and safeguards for continued growth and anti-financial plundering strategies. Explaining the ins and outs of banking structures and real estate asset refinancing power for increased wealth, this large project is a cake walk.

Let me show you what selling $10 Trillion of ZEGE EMI PENV ON-SITE energy systems looks like.

DO YOU FEEL LUCKY?
CAN YOU DANCE?
Let's dance.

GROCERY CHAINS use major mega electricity: expensive!

ZEGE ON SITE energy systems

Go forth and save them $ billions.

Each site will need a 50MWe-100MWe ZEGE EMI PENV ON-SITE power plant sale of $60-$120 million at each store to cut their electricity bills to 2 cents/KWe.

At Wal-Mart that would amount to 80% in savings…$240 billion I estimate. Like shooting…who would put monkeys in a barrel, anyways.

Grocery Chains have 1-3,000 stores each and pharmacy chains over 5,000, **Walgreen** has 7,000. Sell the whole barrel of **Walgreen** chain zebras in one shot.

Will ZEGE EMI PENV sell itself? Just ask Corporate CEO's or CFO's

"What are your energy bill totals, nationally?"

ENERGY SALES IS ALL ABOUT KNOWING SOME SIMPLE MATH.

Know it and you control the energy cost sales conversation. You can't sell unless you do the math. Once you understand how to calculate you have the exact cost of any business and you can intelligently tell them just how much ZEGE EMI PENV will save them or just say 2 cents per KWe an hour.

I estimate the energy costs at

7,000 store Walgreen $70 billion?

A 1,000 store Grocery chain $20 billion?

Wal-Mart Sam's Club $30 billion?

Let's do the math and see if I am close.

Let's do the ZEGE energy cost estimating math.

There are 8,700 hours in a year.

 A Kwe costs on average 10-12 cents.

2,000 store Home Depot, 2,000 stores Lowes, 600 locations COSTCO,

30Mwe each store per hour of electricity purchases.

In 1 Mega watt there are 1 million watts

In 1 million watts there are 1,000 KWe.

50MWE -100MWE divided by 1 thousand, comes to 50,000KWe-100,000KWe/ every hour of every day.

So divide by 10 cents and you get the hourly monetary cost of each store's energy on average.

There are 8,700 hrs in a year, 24hrs in a day

So we now know the cost per hr. for our systems we are trying to sell to Corporate CEO and CFO who have already done the math in their head is $5,000-$10,000 per hr. $120,000-$240,000 per day,

$43.8 million -$87.6 million annually per store

Redoing the math for only 30MWe sized units below $40 million per store cost Times 600-2,000 stores

Totaling for 1,000 Home Depot stores below $40 billion cost. Excess electricity sales will significantly cut energy costs and help pay for the installations in a few years time, leaving a 2 cent/KWe Independent ON-SITE electricity energy operation cost.

LOWES we estimate 1,000 stores at the same $40 billion cost

COSTCO 600 stores $24 billion

Phoenix can let us us know when they are going to make different sizes for NOW it is 30MWe as the smallest and they combine 100MWe units at any location easily, which is the largest they are having manufactured.

OR you could just ask the Corporate Executive representing them what their annual energy bill is.

NOTE on sales unit energy need sizing.

Most box stores under further review can be serviced sufficiently with 30MWe XEGE-PENV-EMI electricity power plant units $40 million eacc,

I am just estimating. Am I feeling feverous or what? Grocery stores are usually so cold. ZEGE can temper those energy bills and Global warming too. Can someone call Dr. House? Maybe cut the bright lights with SKY Lights. Or not! We are ready with National financing.

LET'S MOVE ON.

ZEGE-emi-ON-SITE

SHOP Til You DROP MALLS, Trains, Planes and Big Red Rigs.

ZEGE-TRUCK

ZEGE Inc owned 300,000 Electric and Hydrogen fueled Trucks & MEGA Truck Stops

I hear tell, Hydrogen fueling a big rig at ZEGE Comfort Trucking Depots, will take less that 10 minutes, and Interstate Truck Stops Motels and Restaurants with a side of fries. Oh My! ZEGE wants a major do-over of 50 interstate truck stops. Do it ALL OVER by spending more than over with EMI PENV tech. Plug your electric Peterbuilt in to charge and get some vittles and sleep, dinners on Dr. House.

Freightliner of Savannah

ZEGE plans to write a purchase agreement for 300,000 Big Rigs, half ZEGE EMI PENV 30MWe units powered and the other half using ZEGE EMI PENV power plant manufactured hydrogen fuel.

Let's look at the financials and the scope of this national ZEGE EMI PENV project of $5.5 trillion upfront Loan. This project produces $500 billion annually in profits, Generates $1.3 trillion annual gross revenue $13 trillion in 10 years, with $800 billion in annual operating costs.

We will retrofit the trucks and retain ownership in Lease to own agreements. When the electric trucks are at truck stops or off the interstate leave we are having the drivers have them kept running and plugged into the electric grid to generate extra income. We will tell all the drivers to do the same when the trucks are not on routes.

The driver/owner operator can get credit for their own home heating, hot water and electric bills, charge or hydrogen fuel their family's cars and pocket the difference when the Grid sends them a check each month for excess electricity.

They run fuel free no cost technology so the independent or ZEGE drivers will save $60,000 in diesel fuel each year and they can take that to the bank.

Lets lease-share the trucks Pony Express style to keep them running 16 hours a day. Let each driver sleep in the truck, his home or free in our motels at our multi- clean energy fueling ZEGE Mega Truck Stops.

WE will sell only our own manufactured clean renewable fuels: hydrogen, electric, and super clean syntech diesel, we don't sell gasoline, it smells something awful anyways.

Also order another 50 complete Truck comfort stops and place them 500 miles apart because our trucks run two drivers and need little fuel or none to operate. We manufacture our own fuels on site. Cost for these localized decentralized Interstate Truck and Comfort stops trucks and drivers, ZEGE power plant and electrolysis hydrogen fuel

production on site is…

NOTE the smallest Phoenix Energy EMI units are 30MWE and way too powerful for the needs of one single truck so we are thinking to run all our fleet trucks in Convoys with a trooly like third rail connection truck to truck with mega batteries on board to allow independent running to drop load go to a ZEGE MEGA TRUCK STOP or for any reason the driver wants to run independent. Any truck can recharge by re-connecting to any Tractor Trailer ZEGE Power Plant electric supply unit while running at 50-70 MPH.

The Convoy Charging tractor Trailer will store excess electricity by converting to hydrogen with an electrolysis system: water and hydrogen storage units will be in the trailer.

150,000 ZEGE powered electric unlimited no fuel trucks $3Trillion lease to own or a self funded profit generating project.
Lease per yr. $6 million my cost $2 million/ Gross rev profit $4 million times fleet $600 billion.

     Loan cost $5 trillion gross rev $6 trillion. Gross profit $1 trillion.
Per hydrogen truck lease $300,000yr. 150,000 trucks gross lease rev. $375 billion ten year $3 million each truck. 150,000 total gross lease $450 billion minus cost $150 billing net gross rev $300 billion. Per truck 50 truck stop income $6 billion.

150,000 hydrogen fueled trucks $1 million each, cost $150 billion total. A self funded $300 billion profit making project.

The ON SITE ZEGE 1,000 Mwe electricity and fuel manufacturing power plant can feed thru the grid operations by agreement to 30,000 homes 10 cent/Kwe electricity to pay for the entire Interstate ZEGE comfort truck fueling Stop. Electricity to homes Revenues $60 million annual.

Fuel revenues 350,000 trucks a year, 1,000 a day av. $100 yr. gr sales $35 million.

30,000 cars a week, 1.5 million a year average sale/car $20..

Total annual sales gross…$30 million.

Restaurant sales $20 million.

Motel sales 130,000 room nights/$50 …total gross $65 million

Fuel sales and customer sales are ongoing profit based operations around the world and we will make more because we are owning trucks, making our own fuels cheaply and selling excess power for multiple homeowner uses transportation, electricity, heat and hot water.

Each of the 50 national Truck stops will cost $2 billion, be self funding generating daily multiple money streams totaling $26.15 billion per stop of the 50 Truck Stop Mega operations or $1.3Trillion gross profits annual, Over ten years $13 trillion

Loans of $5.5 Trillion total to start.

$13T Gross rev minus Loan plus interest $7.5Trillion= Gross profits $5.5 Trillion over ten years. Annual profits $500 billion/ year.

ZEGE-Hawaii Power
$6 billion

This project is very important in part because HEI/HECO has deciphered thoroughly the Phoenix Energy technology and the fact that they can totally reject it even though the entire decision making officials have told them they must select 100% clean renewable energy to replace oil and coal and they are defiant, Having rejected the PUC's, Legislature's and Governor's directive. It is a wonder that they have not stood in the Hawaiian Palace and screamed FUCK Hawaii and Fuck you all to hell we are Big Oil and Gas.

LET the no controlling authority deregulation of the Hawaii small clean energy providers begin. We intend to sell energy systems using PENV EMI ZEGE technology to make the old fossil fuel centralized expensive and dirty electricity power plants and electric grid irrelevant while letting it continue to operate at their PUC authorized profit structure, As people decide to use our systems in many ways the old HECO grid shall just not be needed or used in 2 short years. HECO's fossil fuel system is dying and running itself into the ground. We will even pay them out of pocket so the last few customers shall not have to take all the financial burden of its dying days alone, unless, HEI/HECO, wants to join us, which we would happily oblige.

ZEGE President has recently asked First Hawaiian Bank, CEO Robert Harrison and Bank of Hawaii CEO Peter Ho to not only fund the ON-SITE independent power plants but to install a unit in each of their Corporate Honolulu Office Tower Headquarters. XEGE President, Jeffrey Barrett has sized twin 30MWe ZEGE PENV EMI units in the proposed installations each of the 30MWe mini power plants is capable of carrying the entire building energy load, even when one unit is down for maintenance.

The excess electricity production shall be sold HECO for a profit and the elimination of the two banks energy need costs. Current electricity costs for the each building is estimated to be $12 million. Profits from selling excess electricity would be beyond free electricity over $20 million per year including financing costs.

System financing to encourage duplication throughout Hawaii, Honolulu and Waikiki is for a 10 year loan of equal annual money streams from half of the excess electricity cash generated from sales to HECO. Zege invites all Office tower management, residential and Hotels to call him at ZEGE LLC 808-397-0344 for a similar installations sales meetings for review of their own ON-SITE mini power plant system that eliminates your building's electricity bill and pays you excess electricity production cash from a HECO contract.

ZEGE emi Hawaii POWER 2 year goal 2 cents/KWE for ON-SITE users…10 cents/KWe for non electricity producing customers.

ZEGE-emi Hawaii Power Electric underground smart super electricity conductive Graphine in localized systems with 720 degrees interconnectivity. A hydrogen trunk line shall connect all new underground Graphine mini grid systems. This is the face of the new grid with back up of hydrogen gas under pressure piped by trunk lines north to south and east to west from many new EMI ZEGE PENV power plants'

ZEGE-emi Hawaii Power has loans pending submission for the entire project of $10 billion.

We will install a completely new 720 degrees instead of the usual one way electric grid. Electricity will flow in this super smart grid system in all directions giving and taking only 100% clean energy at 2cent/KWE to 8cent/Kwe in an underground conduit housing threaded strands of ambient temperature super conductive Graphine in select parts of Hawaii where energy needs especially electricity are most in demand.

ZEGE emi Hawaii Power with Hawaiian Gas will manufacture and offer to deliver to the current HECO plants, new electric and hydrogen fueling stations, and many businesses throughout the island. $3 billion will be dedicated to build and operate all new decentralized power plants using EMI ZEGE PENV technology systems

ZEGE emi Hawaii Power will manufacture hydrogen and electricity from these plants.

ZEGE emi Hydrogen energy backup storage and distribution system Excess electricity shall be converted to hydrogen and in high demand reconverted to on demand electricity coordinated by the interconnectivity of the new Graphine smart grid and 10,000 PENV EMI ZEGE 30MWe-100MWe ON-SITE electricity generating power units costing ZEGE $3 billion.

ZEGE-emi Hawaii POWER ON-SITE small independent 1-60 renewable power plants, will be fully paid for by PPA lease or sales agreements and interconnected to the new underground Graphine super smart super conductive grid. These all new ON-SITE electric plants shall be located at businesses located in Honolulu, Waikiki, and Hawai-Kai, and again in Pearl Ridge Pearl City, Salt lake Kenehue, Kapolei, the North Shore and Waianae or other clean fuels such as hydrogen based diesel syntech and jet fuel syntec.

ZEGE-emi Hawaii Piped and Stored Hydrogen will be installed by ZEGE and Hawaii Gas throughout the island in a partnership agreement we seek. The hydrogen systems will be fully paid for by ZEGE, LLC. The new hydrogen system is an integral necessity to achieving grid reliability and in lowering energy costs by integrating all energy needs an a synergistic system. Hydrogen and home capacitor and small $500 home batteries in every home will even out the grid demands for power. Hydrogen shall be the main energy storage technology of this system because it is so easily converted to many types of energy uses. It is one of the many grid backup systems. This interconnectivity allows conversion of energy back and forth for multiple uses such as Sewer treatment. Desalination, transportation fuels and airline jet fuel.

ZEGE emi Hawaii POWER ON=SITE We are offering ON-SITE ZEGE EMI PENV systems demonstrate the technology for free to Bank of Hawaii and First Hawaiian Bank whom we shall request to fund this $6 billion zero emission project

ZEGE-POWER-GRAPHINE

ZEGE-emi-Power superconductive, super integrated GraphineGRID

An INTEGRATED Smart Superconductive Grid is a comprehensive zero emission energy manufacturing, transferring and energy utilization and storage system with underground super smart super conductive 720 degree integrated electricity and hydrogen systems for cities and suburbs. They will be sold worldwide in partnership with ZEGE-PENV-EMI with contracts for Mitsubishi International Corp. manufacturing. Using EMI-Phoenix tech, Hawaii Gas underground piping systems or new main hydrogen pipelines and all ON-SITE independent power producers we will concentrate on manufacturing hydrogen fuel from central and ON-SITE energy producers who will be integrated to the graphine underground localized grid and or hydrogen gas pipe line. Both energies will be interchangeable as demand requires. We can produce all transportation fuels and cross connect the independent electricity providers and the decentralized interconnected ZEGE electri-hydrogen producing plants. Hydrogen underground piping shall be the backbone of the NEW Grid.

Hydrogen is our energy storage and distribution solution. We invite the old fossil fuel powered plant or grid operators to join our integration technology. Transportation energy demands is twice that of electric needs.

From $10 billion per million population

Super Mega cities $1-$10 Trillion.

10 yr. revenue, serving 1 billion people 500 cities. **$10Trillion.**

GRAPHINE TECH.

Desalination

&

Medical Tech.

References

    5 Hour Energy multi billionaire, CEO Ravi Sajwan has already spun thousands of strands of thin pulled graphine lines as big as fishing lead into thick graphine wire which is a superconductor, These graphine wires are so efficient with no energy loss sideways that they can be used to transfer heat from far underground at high temperatures high enough for high pressure steam and transfer that high heat with zero losses to the surface. No heat loss. You'd think he was Phoenix Principal Michael Dooley's lost twin.

Ravi Sajwan is giving the world this technology for free. The most important thing about his commercially viable process of spinning graphine wire, is that graphine is a superconductor with very little heat gain, which allows it to handle far greater amounts of electricity loads than copper high voltage wire now in use.

ZEGE will be in direct contact with the **Billions in Change research labs** to get greater manufacturing processes from **CEO Ravi Sajwan** whose team has done a stellar processes advances that anyone to date. By funding this research further in his labs and our manufacturers especially Mitsubishi International Corporation, we ensure a great super smart 720 degree smart grid product to date is not even on anyone's installation manifest. This technology allows us to install a super integrated smart underground conduit 720 degree localized power grid. Electricity can be pushed beyond the one way power line limitation of the past century. Anyone producing independent ON-SITE electric clean power who has a need for more than they are producing can draw down, and if they have excess they can send it across this new smart integrated grid for profit or banking in credit, cash, for hydrogen storage or for sale.

This is, key to selling huge ZEGE EMI PENV POWER systems. This will cascade around the world the buyout or easy replacement of all fossil power plants and old style very limited electric grids. I assume most people know that All Electric Power producing owners and grid operators are associated EDISON electrical engineering club. Many of the electric company owners are also Federal Banking associate Board members. Some sit on the all sit on the Presidential Energy Reliability Commission as well Home Land consultants to terrorism defense.

All seemingly, well intentioned, but under deregulations the big MA BELL and Edison National and Regional electric Companies were banned from owning everything. But under these associations secret meetings still run the show. Stock ownership replaced big corporate control. Many stock owners own many power company stocks and sit on multiple boards.

They will come to join us in this new renewable energy transformation or get run over having blinders, asleep at the helm. Fossil fuel investment and reliance is the energy Titanic for them. Just try and sell them new robust 2 cent/KWe electricity and watch how they circle the wagons to deny you access to their money making machine.

Try to purchase several systems and then convert them to 100% clean energy and then go on rolling up one company after another returning it back to clean energy local alliances. Or simply buyout the entire stock and take over their company right out from under them and go green overnight now that ZEGE-Mitsubishi-EMI-PENV are able to produce 2 cent/Kwe electricity. Or build an underground localized new clean energy cooperative system with all the new tech and see them cry. They are dirty fossil fuel importers and consumers, ZEGE-EMI-Phoenix Energy of Nevada are energy producers. They cannot possibly compete with that fact. You are paying for their mistakes. Buy from us and put a unit ON-SITE and make cheap electricity yourself. It is that easy.

You can call me to answer any questions or to order a ZEGE-EMI system for your company. US STEEL divested of oil to heat their iron and coke processes 70 years ago to use Electromagnetic Induction heating at tenth the cost, isn't it about time for the rest of us to catch up to them? I am talking to every nation and Corporation, University, railroad or airline. Call me today. ZEGE Inc is a Global Company.

Zero Emission Global Energy, LLC

Jeffrey Barrett, President

cell 808-397-0344 Hawaii

Puc.report@gmail.com

Limitless Energy

Energy

Energy straight from the core.

We can power the world with clean resources that are right under our feet.

Limited Clean Energy

Burning fossil fuels and creating nuclear reactions to generate electricity comes at a high cost – politically and environmentally. Alternative methods are limited and come with tradeoffs. The answer to these problems is right under our feet.

THE SOLUTION

Our Solution: Limitless Energy

Not too far below the surface of the Earth, it's hot. That heat can create enough clean energy to power the world, and help keep things cool above. Using cables made from graphene, a form of pure carbon 100 times stronger that steel, that heat can be conducted to the surface of the Earth to run turbines and generate electricity – without burning anything. We call it Limitless Energy.

EMI-ZEGE LLC

EMI Zero Emission Global Energy president Jeffrey Barrett wants to personally express individually to **Phoenix Energy of Nevada** Principals: Michael Dooley and Brian Smith, thank you gentlemen for our fast and furious talks and emails back and forth over these last months it is truly a privilege to know you have given back to everyone on this planet by bringing YOUR vision to reality. Providence is applauding you today I am told. I am in your dept and you have my deepest thanks and respect. and

And to MITSUBISHI International Corporation President and CEO Mr. Takehiko Kakiuchi and to all our Banking Partners we are eager to personally to meet you and ask that you and your Corporate family become our friends and business partners in these Global undertakings. You have by decades of effort, with thousands upon thousands of the most dynamic people created dynamic universally respected $Trillion enterprises so rich with so many undertakings, beyond our grasp to understand, the orchestration of such earnest complexity in the past hundred years, is breathtaking. We need for you to have these capabilities, because our needs are of the same dynamics scope. Our success depends on you championing our **Cause** because it is your own **Cause** as well I suspect.

We need you.

You need us more.

I am baffled how Japan and Germany sit with such capabilities yet, with all this ability and yet have become 97% dependent on dirty imported expensive foreign fuels. I am chagrinned the whole world as well cries out for a solution more impactful to come in like a raging river to overturn this situation. You need a Champion to take up the task. You need a David to this Goliath dilemma. ZEGE and PNEV have this Goliath by the throat.

Yes, we be few, Yes, we are bold braggarts. WE do not come begging we bring proven tested works which are known and respected in the highest chambers of the world. Leaders seek our knowledge and have used it to advantage. They can sign all the proclamations they want. They can fund infrastructure and float bonds but without us they are impudent to this dilemma. I told China fund AIIB at 4 Quadrillion. They half fainted $50 billion they said. Then they got up braced themselves and went about to Britain $50 billion over 10 years and again they took courage and funded France and Germany twice again that amount so far less than a trillion in funding. I have asked and do ask still for their financial due diligence. I wrote a good economic plan for them and India so I expect respect and a better reply of my requests for financing,

I will not beg. I will do what they refuse, myself. I have created wealth for them and they deny me in my hour of need. You have wealth but lack an economic powerful manufacturing vision to fuel your citizens need to work or go broke suffocating in your piles of cash. Phoenix Energy has done the same dance before power brokers on high for 4 years now and been two faced told great stuff, bloke.

They go empty handed with empty promises. We are not amused and we are not silent. We are not powerless like you.

We have slain many Lions and Bears so have no fear. Take up the sword with us.

Look at ZEGE solutions. Look at Phoenix Energy of Nevada's technology.
We are the champions to this Cause. Join us. I could ask the Deputy Crown Prince of Saudi Arabia for the money. I could convince him to, alone, fund our needs, but then you would have no bold manufacturing enterprise to keep meeting your manufacturing needs vision.

And no large enterprise to feed your wonderful enterprises and industries and utilize all your flush with cash Financial Corporate skills so your people may stay productive. Without a growth project continuing as with China, we look to see FAILURE. Japan and China face a precipice as the global bankers want to do from time to time plunder both your nation's wealth as they have done in the recent past: first Mexico, then South America and then the entire Pacific Rim.

In 2008, it is as if they cannibalized their own children, when they orchestrated the plundering and bankrupting of hundreds of the world's largest manufacturing Corporations, Global Re-Financing and many major Financial Institutions and Banking system world wide. Whe are these Pirates to our desires for prosperity? You have your names and I have mine. I watch them closely and tell them God and I are determined to bankrupt you.

The oil price drop is manipulation par excellence to plunder many nations today. This is no game they play. We are wise as well. Plundering China is in the wind. I built in financial protections for them against such events. They will not be plundered and why should any reasonable man want to do that to these so wonderfully industrious and enterprising hard working people unless they were wicked. WE have known many of these people, they know us, we will win our Cause, with or without them. Already they are coming over with enthusiasm to fund Renewable energy. It is needed and they now believe. Warren Buffett, David Rockefeller and Bill Gates et al have divested of oil positions. Huge Financial positions. Now even Deputy Crown Prince Mohammed Bin Salman the new Saudi Arabia Financial Minister has made dramatic unbelievable moves for Saudi Arabia who has decided his country will come to bankruptcy and chaos if he does not act boldly and he has saying of their new Vision 2030 we need to loose ourselves from this fossil fuel devil's noose. They are divesting themselves from oil too about to sell 5% of ARAMCO in a $2-$3 trillion IPO as you know, setting up the largest Global Sovereign Fund in the world to date.

Mohammed Bin Salman is acting. Japan has acted in its own Vision 2020 electric car and hydrogen system technology and now Germany has acted with a Russian agreement to a long term low priced LNG deal has ated, to break free of their own, long time 40 cents/KWe fossil fuel devil. China in funding Britain $150 billion, Germany and France another $150 billion apiece plus internal spending of over several $trillions toward clean energy and energy infrastructure over the next ten years.

I humbly and boldly ask for your help in bringing Fuelless technology from Phoenix Energy embedded Electromagnetic Induction technology to the world and very many transportation and energy producing systems.

I am making a public appeal the the Saudi Arabian Finance Minister Deputy Crown Prince Mohammed Bin Salman. In my letter thru the Saudi Arabian Washington Ambassador I told you of the healing God told Jordan King Hussein he was doing in him to cure him totally of cancer. I reviewed the events soon after when King Hussein went back from his long medical stay in US Minnesota. On his very public return to Jordan the King addressing his people and the world did not say the God of Jeffrey Barrett who brought me the healing message and of Jesus and before that Moses and the Jews has healed me from cancer after a long terminal watch. I thank this, The Living God, father of His holy son Jesus. NO he said ALLAH has healed me. Some would say what is the difference between a God who wants a robust dialogue with those who seek and speak to him expecting a replay of all his created people on the planet to change them into his image and Allah of whom the Prophet Mohammed said of himself talking to Allah, I am the last one allowed to talk to God.

Thus Mohammed was Silencing God to many billions over the centuries in the Muslim World. I say Mohammed Bin Salman tell the Muslim world God has given you and your Father two dreams for the saving of the people of your nation and the world from certain death and nuclear destruction from these evil men. God says it is time for Judgment of the world. Tell them I told you God the Father of Jesus whom came down manifesting God through his flesh to the world in Spirit 2,000 years ago and thru many others ever since has spoken to Jeffrey Barrett to tell you Prince Salman, I myself already have been shown by God your two visions and how to implement this great dreams message of how we must all prepare to enable billions to survive this GREAT DAY of the LORD. . .

Tell the world please Prince Salman what the dreams are and that God of Jesus not Allah has spoken to you and the Prince of the work ahead needed to be swiftly undertaken and completed over three years only before the nuclear global war. Tell your Muslim brothers God the Father can be approached in dialogue and spirit to change us into his image as Jesus came to show how starting at Pentecost for us. God The Promise come to live inside mankind as his children once again made thry trials into new creatures not by rote prayer and formality but by a robust dialogue till we be made perfect in the image of the Son of the Living God manifesting his presence by whom alone can save.

Remember what happen after King Hussein denied hearing and being healed by this the Living God of Jesus and not Allah of Mohammed. For that God struck him dead in a month. For unthankfulness and lie to the Muslim world and to rejecting God's desire to heal him for the express purpose of telling the Muslim world dialogue with me. The Pope is similar if his fraud saying he knows God and represents him.

He is in League with the US and other nations to nuke the world so he can remain the only religious ruler as long as the Jews be blasted off the earth lest they come back thru Christ to know God again and re-establish their leadership in loving God which like Herod trying to kill baby Jesus killed 1,000s of babies. Hitler and the Pope killed as many of the Jews as possible for this purpose of the Vatican and now he is at the UN, White House and US Capitol making speeches promoted before the world as their religious go to guy. He has sanctioned the nuclear war. For this Pope Benedict resigned and is now kidnapped incommunicado in Rome by the Cardinals.

He had listened to God who said shut this fraud thing down, you are my main enemy for 2,000 years. Note this Salman Britain, Germany and France have pledged not to come to the US lead global nuclear war. Saudi Arabia thru you and your good father must answer God by opening up the Muslim world to start loving the rest of the world and let religious freedoms run through the Muslim world again. I am ready to interpret the robustness of the dreams that are designed to build Saudi Arabia into the largest farming and cattle country in the world saving seed and foods for the world enough for all that survive the nuclear war to live after that chaos.

All trees and plants will not survive. Save enough seeds to replenish the planet fully. Carve ten shipping canals of sea water deep into the desert creating a super sea then build new clean energy suburban clean energy communities and come and go around the world with new cargo ships and planes buying farms everywhere and buying foods from every nation for the preparations needed farming and storing enough provender for 20 years to feed the rest of the world. Saudi Arabia will be The Living God's Seed and Food Arc for the world to survive.

Build towers of super cooled sea water inside stainless steel condensing fresh water producing desalination towers to water the entire Middle east. Use the new ZEGE self sustaining clean cheap zero emission PENV EMI technology and help fund it t be implemented around the world so after the refineries and global infrastructures are totally destroyed your help in establishing models and many of your own and globally funded decentralized local communities with self sustaining clean energy will survive intact. All other central systems will be destroyed. Space communications shall be decimated. Build local wifi etc.

Hurry my friend. Listen to God he has spoken to many and me and now to you and your nation and the Muslim world I beg you do not say he has not or Allah spoke to me. For your life in is God;s hands this day, please answer him here I am what shall I do, and do it.

Jeffrey Barrett, president EMI ZEGE, Inc.

520 Pine Street #318

Wahiawa, Hawaii 96787

808-397-0344

Puc.report@gmail.com

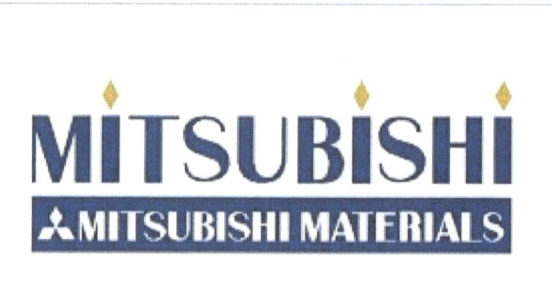

Message from President and CEO

MITSUBISHI RESEARCH INSTITUTE TO HELP MONITOR CARBON ABSORPTION BY FORESTS

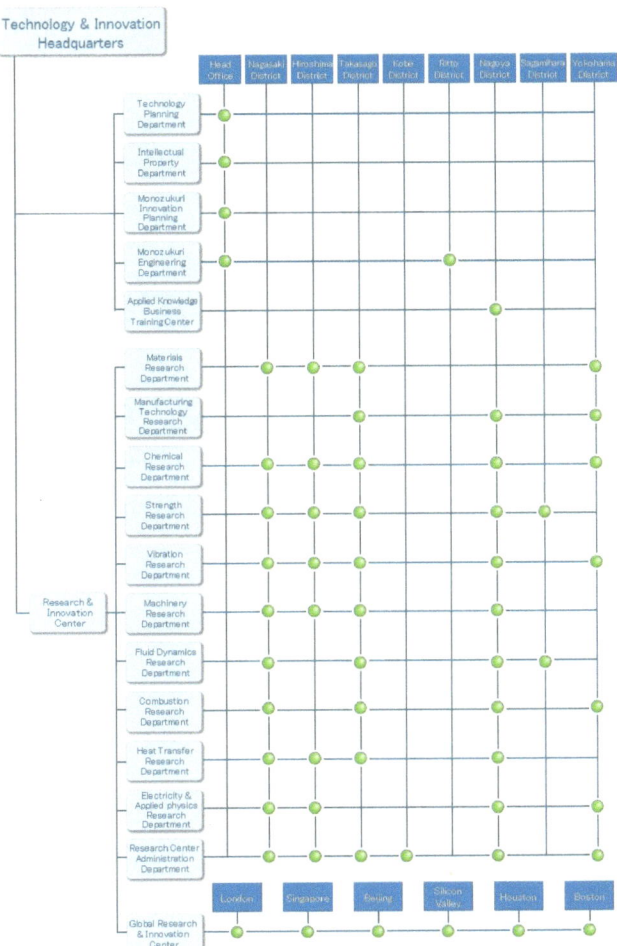

 三菱東京UFJ銀行  三井住友銀行 SMBC SUMITOMO MITSUI BANKING CORPORATION

 三菱東京UFJ銀行

First Hawaiian Bank

MITSUBISHI HITACHI POWER SYSTEMS

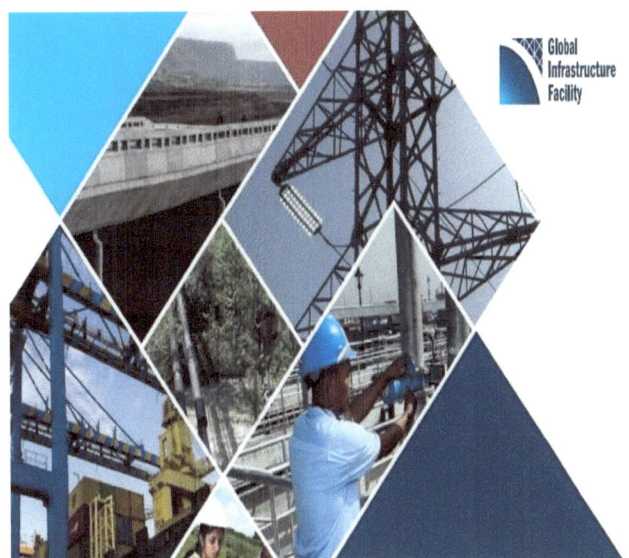

www.ingramcontent.com/pod-product-compliance
Lightning Source LLC
Chambersburg PA
CBHW050721180526
45159CB00003B/1097